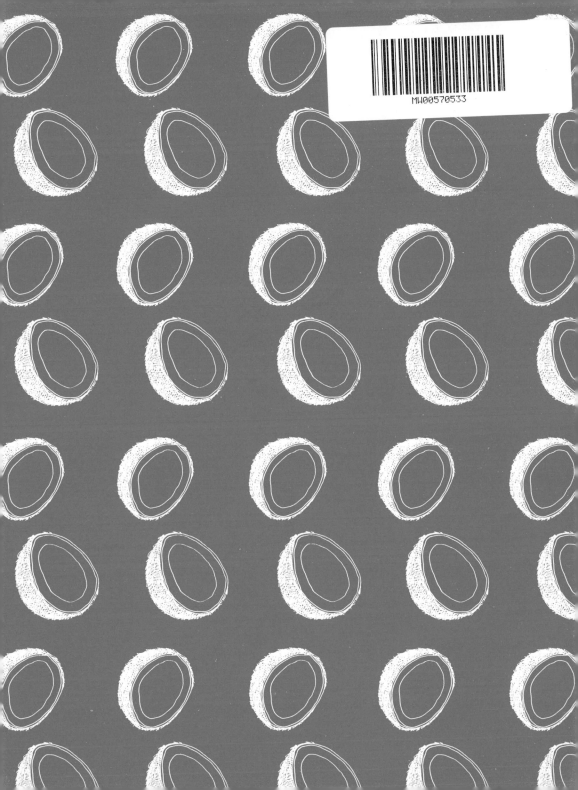

COCONUT

COCONUT

40 IRRESISTIBLE ENERGY-PACKED RECIPES

EMILY JONZEN

PHOTOGRAPHY BY CLARE WINFIELD

KYLE BOOKS

CONTENTS

6.
INTRODUCTION

12.
MILK & WATER

34.
OIL

58.
FLOUR

74.
DRIED

92.
INDEX

COCONUT IS GREAT

The term "superfood" has been thrown around a lot in recent years, referring to numerous ingredients that some home cooks will never have heard of. New scientific research instructing us what to eat and what to avoid can be nothing short of confusing.

The myriad health benefits of the humble coconut, however, cement it as a superfood that is here to stay. Coconut, in its many forms, has been proven to be rich in vitamins B, C, and E, as well as the minerals selenium, iron, calcium, magnesium, and phosphorous. Coconut milk is a great alternative for those with dairy intolerances, while coconut water is so high in phosphates and natural electrolytes that it has become the recovery drink of choice for athletes. Coconut flour is also a great source of fiber and protein. Far from an inferior substitute to dairy and gluten, coconut packs in delicious flavor with added nutritional punch.

THE BENEFITS OF COCONUT

Infinitely versatile, coconut can be used in simple, everyday recipes to delicious effect. This book is divided into four chapters by ingredient. The Milk & Water chapter provides easy recipes for smoothies and juices, soups, and curries, while the Oil section focuses on everyday meals, from granola to risotto. The Flour chapter offers recipes for delicious cakes, pancakes, and bread and the Dried section shows you how to enhance dishes with the earthy nuttiness of coconut.

BUYING THE BEST

Coconut products may be more expensive than their mainstream alternatives, but a little goes a long way, so buy the best you can (see pages 10 to 11).

Top tips for buying good coconut
♦ When buying a fresh, whole coconut, check for ripeness by shaking it to make sure there is still plenty of water inside—a dry coconut will be overripe and soapy-tasting.
♦ Check that there are no cracks in the skin and the dark brown "eyes" at either end are intact and dry.

How to open a coconut
To open a coconut, place it on a sturdy cutting board with the "eyes" facing up. Determine the softest eye by holding a screwdriver or skewer to each eye and seeing which pierces most easily. Pierce the eye, then drain the juice into a pitcher.

Now turn the coconut on its side and use the blunt edge of a heavy knife to repeatedly tap around the circumference of the coconut until it begins to split.

Continue until you have two equal parts. The flesh can now be eased away from the shell with a spoon.

How to store a coconut
A whole coconut can be refrigerated for up to two months. Once opened, the flesh can be kept in the fridge for up to three days, covered. Alternatively, the flesh can be grated and frozen for up to six months.

SOURCING

Purchasing coconut in its many forms can be a little pricey so it's important to make sure that you get the best value for your money.

Water

Found inside the coconut, its sweetness differs depending on age and environment. High in electrolytes, fiber, potassium, magnesium, and phosphorous, it is ideal to drink after exercise. Because of its balanced sugar levels and vitamin and mineral richness, it is far healthier than high-sugar fruit juices. While the water can be enjoyed alone or in drinks, it is not economical to regularly drain coconuts. Happily, it is readily available in cartons. Try to buy unsweetened, and organic for purity. Once opened, refrigerate and use within three days.

Milk & Cream

Made from grated coconut flesh soaked in hot water. Cream rises to the surface and is skimmed off, then the remaining liquid and flesh are squeezed through a cheesecloth to extract the thick milk. A great alternative to dairy, as it is lactose-free, hormone-free, and rich in B vitamins, iron, and copper.

Coconut milk and cream also come in cans and cartons and can be found in most supermarkets. If you are concerned about the fat content, opt for full-fat coconut milk and halve the quantity; reduced-fat milk has been watered down. Keep unused milk or cream in an airtight container and refrigerate for up to three days.

Flour

Made from dehydrated, ground coconut flesh. It is gluten-free, has almost double the fiber of wheat bran, and is far richer in protein than wheat, rye, or cornstarch. It is very light, making it useful in cakes, muffins, and pancakes, but it may need to be mixed with xanthum gum to mimic the stretchy gluten proteins that hold wheat flour products together. It is very absorbent and swells upon contact with liquid, so ratios to egg and other liquids are different to wheat flour.

Coconut flour is available in large supermarkets, health food stores, and online. Try to buy organic; it is fairly expensive, but you will use far less of it than conventional flour—often 3 to 4 tablespoons instead of 1 cup. Once opened, keep in an airtight container in a cool, dark place until the expiration date.

Dried and Desiccated

Made by cutting or grating the flesh, blanching it to remove impurities, and then drying until the moisture level reaches 3 percent (from 19 percent). Try to buy organic, and sprinkle it over smoothies, fruit, granola, and yogurt, as well as curries, chicken, and other savory dishes. It is delicious raw or lightly toasted and will keep in an airtight container for several months.

Sugar

Boiled and dehydrated sap of the coconut palm flower, it can be used in place of sugar, (though it is richer in taste and lighter in weight, so don't substitute it measure for measure). It has a fairly low glycemic index, which makes it a popular substitute for diabetics. Some manufacturers mix it with cane sugar, so check the label to ensure it is pure.

Oil

Coconut oil can be swapped for olive and sunflower oils in most cooking. It is great for sautéing and gentle cooking, but its relatively low smoke point (due to it being unrefined) means it is not suitable for deep frying or cooking at very high temperatures. It has a mildly sweet, nutty taste that complements lots of foods. You can buy unflavored oil, but it will be more processed and less nutritious.

While coconut oil is high in saturated fat, it is beneficial for the heart due to the high percentage of lauric acid, which prevents high blood pressure and lowers cholesterol. Its antifungal, antibacterial, and antiviral properties are also believed to strengthen the immune system.

The myriad health benefits of coconut oil will be most potent in unrefined, organic, virgin, and raw coconut oil, free of the preserving chemicals that may be found in cheaper, refined coconut oil. Good-quality coconut oil is now available in most supermarkets as well as health food stores. The oil is solid at room temperature and will need to be melted over low heat as necessary. Store in the jar with the lid firmly sealed and away from direct sunlight.

MILK
& WATER

COCONUT & ALMOND GREEN SMOOTHIE
*VEGETARIAN *DAIRY-FREE *GLUTEN-FREE

The creamy nut butter and sweet coconut water counterbalance any bitterness from the greens in this refreshing smoothie—a perfectly nutritious start to the day.

Serves 2

2 bananas, coarsely
   chopped and frozen
2 tablespoons nut butter
A large handful of spinach
2 cups coconut water
Ice cubes

1. Place the ingredients in a blender and process until smooth. Serve immediately or pour into a portable cup to drink on the go.

ORANGE, GINGER & COCONUT JUICE
*VEGETARIAN *DAIRY-FREE *GLUTEN-FREE

This warming, spicy drink is ideal for fighting off winter colds and flu. Coconut water is anti-bacterial and anti-fungal, whilst the ginger is also anti-viral.

Serves 2

2 oranges, peeled
1½-inch piece of fresh
   ginger, peeled
2 cups coconut water
Crushed ice, to serve

1. Feed the ingredients through a juicer and serve immediately over crushed ice. If you don't have a juicer, simply squeeze the orange into a glass and finely grate in the ginger. Add the coconut water, stir, and pour into another glass over ice.

CARDAMOM HOT CHOCOLATE
*VEGETARIAN *GLUTEN-FREE

The creaminess of coconut milk makes it ideal for a luxurious hot chocolate, plus it has more iron and copper than cow's milk, to help keep your immune system healthy. Cozy up with this decadent, aromatic treat.

Serves 2

1 cup coconut milk
1 cup boiling water
2 green cardamom pods, split with the back of a knife
3½ ounces good-quality dark chocolate, minimum 70 percent cocoa solids, finely chopped
Sea salt
Maple syrup

1. Heat the milk in a small saucepan with the water and the cardamom over low heat, until starting to simmer. Remove from the heat and stir in the chocolate to melt.

2. To serve, add a pinch of salt and the maple syrup, to taste, and reheat if necessary, without boiling. Pour into individual mugs.

This goes really well with the Oat & Sour Cherry Cookies on page 87.

COCONUT & BERRY BIRCHER MUESLI

*VEGETARIAN *DAIRY-FREE

This twist on a Swiss staple soaks overnight in coconut milk and apple juice to be enjoyed with berries and apples the next morning. I've suggested topping the muesli with raspberries, but any fruit works well.

Serves 2

½ cup rolled oats
⅔ cup coconut milk
½ cup apple juice (not
   from concentrate)

To serve
¼ cup whole almonds,
   coarsely chopped
1 sweet apple, coarsely grated
1¼ cups fresh or frozen
   raspberries
Maple syrup (optional)

1. Place the oats in a bowl or glass pitcher and pour in the coconut milk and apple juice. Cover and refrigerate for a minimum of 2 hours, or preferably overnight.

2. To serve, spoon the muesli into two bowls and loosen with a little extra juice, if necessary. Sprinkle over the almonds, apples, and raspberries and serve with a drizzle of maple syrup, if using.

BANANA & BERRY SMOOTHIE BOWL

*VEGETARIAN *DAIRY-FREE *GLUTEN-FREE

Start the day with a bright and colorful smoothie bowl. It's a delicious way to pack in lots of nutrients first thing in the morning.

Serves 2

2 bananas, 1 coarsely chopped and frozen, 1 whole
¾ cup coconut milk
1 sweet apple, coarsely chopped
2 tablespoons chia seeds
3 cups mixed berries (e.g. blueberries, raspberries, and strawberries)
Filtered or spring water

To serve
2 tablespoons mixed seeds
2 tablespoons coconut chips

1. Place the frozen banana, coconut milk, apples, and chia seeds in a blender or food processor. Pour in the berries, saving a few for decoration. Blend the mixture until smooth, adding a little water if it is too thick.

2. Divide the smoothie between two bowls and scatter with the remaining berries, mixed seeds, and coconut chips. Slice the remaining banana and arrange around the other toppings before serving.

Opt for frozen berries if fresh are out of season—they're much cheaper and will give the smoothie a thicker, slightly creamy consistency.

SHRIMP TACOS WITH COCONUT & LIME

*DAIRY-FREE

These light and flavorful tacos are perfect for a quick and easy lunch or dinner. Avocados compliment the nutritious coconut perfectly, packed with heart-healthy monosaturated fatty acids, potassium, and fiber.

Serves 4

½ cup coconut milk
Juice and zest of 1 lime
A small handful of cilantro stems, finely chopped
7 ounces raw jumbo shrimp
Sea salt
Pinch of red pepper flakes
2 teaspoons coconut oil
4 small corn tortillas

To serve
1 ripe avocado, finely sliced
2 scallions, finely sliced
A small handful of cilantro leaves, coarsely chopped
Tabasco chipotle sauce (optional)

1. First make the dressing: Whisk the coconut milk to smooth out any lumps before stirring in the lime juice and chopped cilantro stems. Set aside.

2. Toss the shrimp with a pinch of salt, the red pepper flakes and lime zest in a nonmetallic bowl. Heat a nonstick frying pan until blisteringly hot and add the oil. Add the shrimp and stir-fry for 1 to 2 minutes, turning occasionally, until opaque and golden. Remove the shrimp from the pan with a slotted spoon and set aside.

3. Wipe the pan clean with a paper towel. Now heat the tortillas in the pan for about 10 seconds, to warm through and soften.

4. Take a tortilla and scatter over a portion of the avocado and scallions, followed by the shrimp. Sprinkle with the cilantro leaves, spoon over the dressing, and top with a dash of Tabasco, if using. Wrap up and serve immediately.

CHICKEN SOUP WITH BABY CORN & LIME
*DAIRY-FREE

This delicately spiced, fragrant soup makes a wonderful meal all year round. It is traditionally fairly spicy—feel free to adjust the chile content according to taste.

Serves 4

1 tablespoon coconut oil
2-inch piece of fresh ginger, peeled and coarsely sliced
4 shallots, peeled and halved
4 kaffir lime leaves, or zest of two limes
1 to 2 green chiles (depending on heat preference), halved lengthwise
2 lemon grass stalks, bruised with the back of a knife
2 whole chicken legs, skin on
5 cups good-quality chicken stock
1 (14-ounce) can coconut milk
2 teaspoons coconut or palm sugar
1 cup baby corn, coarsely sliced
2½ cups button or shiitake mushrooms, sliced
1½ tablespoons fish sauce
Juice of 1 lime, plus extra wedges to serve
A handful of cilantro, coarsely chopped

1. Heat the oil in a large saucepan over medium heat. Add the ginger, shallots, lime leaves or zest, chiles, and lemon grass and cook for 2 to 3 minutes, until everything has a bit of color.

2. Add the chicken and continue to cook for another 3 to 4 minutes, until the chicken is light golden. Pour in the stock and bring the mixture up to a boil. Reduce the heat and simmer for 20 to 25 minutes, until the chicken juices run completely clear. Remove the chicken from the pan and set aside to cool slightly, before stripping the meat from the bone, tearing it into bite-sized pieces.

3. Strain the cooking liquid into another large saucepan, pressing everything down in the strainer to extract as much flavor as possible. Add the coconut milk to the pan, along with the sugar. Heat the soup until simmering, then add the corn and mushrooms and cook for another 2 minutes. Return the chicken to the pan to warm through. Stir in the fish sauce and lime juice.

4. Ladle the soup into individual bowls and serve sprinkled with cilantro and a wedge of lime.

CREAMY RED LENTIL & COCONUT DAHL

*VEGETARIAN *DAIRY-FREE *GLUTEN-FREE

This uses a few spices, but once you've bought them, this is a dish becomes a wonderful store-cupboard dinner. Garlic is a great source of vitamin c and manganese, as well as minerals including calcium and potassium. Ginger is great for blood circulation and has been used for generations as an antiinflammatory and cure for nausea.

Serves 4

1 tablespoon coconut oil
2 garlic cloves, crushed
1-inch piece of fresh ginger, peeled and finely grated
1 green chile, finely chopped
1 teaspoon white mustard seeds
½ teaspoon nigella seeds
1 teaspoon cumin seeds
1 onion, finely sliced
½ teaspoon ground turmeric
1 cup red lentils
1 (14-ounce) can chopped tomatoes
1 (14-ounce) can coconut milk
Sea salt and freshly ground black pepper

To serve
A handful of cilantro leaves
Plain yogurt (optional)

1. Heat the oil in a large frying pan over low to medium heat. Add the garlic, ginger, and chile and cook for a minute or so, until fragrant. Add all of the seeds and cook for another minute, until the spices smell slightly toasted and aromatic. Stir in the onion and cook for another 5 to 6 minutes, until softened. Stir through the turmeric and cook for a final minute, before adding the lentils.

2. Add in the tomatoes and coconut milk and season to taste. Half-fill the tomato can with water and add to the pan. Bring the mixture up to a boil and reduce the heat to a simmer.

Allow the dahl to gently bubble away, stirring from time to time, for 20 to 25 minutes, until the lentils are tender and the sauce has thickened and reduced.

3. Divide the dahl between four bowls and serve with a sprinkle of cilantro and dollop of yogurt. It also goes very well with the Seeded Coconut Flour Bread on page 60.

CAULIFLOWER & GARLIC SOUP

*VEGETARIAN *DAIRY-FREE

Roasting cauliflower and garlic brings out their natural nuttiness and sweetness. This soup is simplicity at its best—few ingredients brought together by the soothing creaminess of coconut milk.

Serves 4

1 garlic bulb, sliced widthwise across the middle
2 tablespoons coconut oil
1¾ pounds cauliflower, cut into florets
1 onion, sliced
Sea salt and freshly ground black pepper
1 cup coconut milk
3 cups good-quality vegetable stock

To serve
A handful of parsley, coarsely chopped

1. Preheat the oven to 400°F. Drizzle each half of garlic with a little of the coconut oil and wrap in foil. Place on a baking sheet and roast in the oven for 20 minutes.

2. Now place the cauliflower and onion in a separate baking dish and pour in the remaining oil. Season lightly and give everything a good stir. Transfer to the oven and roast for 20 to 25 minutes, until golden and just tender. Remove both pans from the oven. By now the garlic should be golden and completely soft.

3. Transfer the cauliflower and onion to a food processor. Squeeze the garlic out of its papery skin and add to the processor, along with the coconut milk. Pulse until smooth, then transfer to a saucepan. Add the stock and heat until just simmering. Taste and check the seasoning.

4. Divide the soup between individual bowls and sprinkle with the parsley.

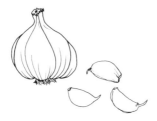

PEACH & VANILLA CHIA PUDDINGS

*VEGETARIAN *DAIRY-FREE *GLUTEN-FREE

Nutrient-packed chia seeds are bolstered by the good fats in coconut milk—a winning combination. These puddings also make a great light breakfast before exercise. Peaches work really well, but the puddings will taste delicious with any fruit.

Serves 2

¼ cup chia seeds
¾ cup coconut milk
¾ cup almond milk
1 vanilla bean split in
   half, seeds scraped out
   with a knife

To serve

1 ripe peach, pitted and
   sliced
A small handful of pistachios
   (about ¾ pound), coarsely
   chopped

1. Place the chia seeds in a large measuring cup or bowl and pour over the milks and vanilla seeds. Stir until everything is well-dispersed and divide the mixture between two bowls or jars. Chill in the fridge for at least two hours or overnight.

2. To serve, top each pudding with the sliced peach and sprinkle with pistachios.

To get the seeds from a vanilla pod, split it in half lengthways with a knife, then scrape the seeds out with the blunt edge of the knife.

COCONUT CREAMS & GRILLED PINEAPPLE

*DAIRY-FREE *GLUTEN-FREE

These little desserts capture the pure essence of coconut. The rich, creamy sweetness is complemented perfectly by the slight tartness of the pineapple. They're easy enough to make midweek for yourself, but impressive enough to serve to dinner guests.

Serves 4

4 gelatin sheets
2½ cups coconut milk
¼ cup coconut sugar
1 vanilla bean, split in half, seeds scraped out with a knife
A few gratings of nutmeg
Half a pineapple, cut lengthwise into eight wedges

1. Drop the gelatin sheets into a bowl of cold water and push down to submerge. Set aside to soak for 5 minutes.

2. In a saucepan, gently heat the coconut milk with the sugar, vanilla bean and seeds, and nutmeg until barely simmering. Remove the vanilla bean from the pan and discard

3. Lift the softened gelatin from the water and stir into the coconut milk mixture until completely dissolved. Pour into four glasses or ramekins and refrigerate for a minimum of 2 hours to set.

4. Preheat a grill pan over medium heat. Cook the pineapple wedges for 2 minutes per side, until lightly caramelized. Set aside to cool slightly.

5. Divide between individual plates and serve alongside the creams.

INSTANT ICE CREAM

*VEGETARIAN *DAIRY-FREE *GLUTEN-FREE

This ice cream really is a kind of magic. Frozen fruit and coconut milk is processed until it whips up into a light, fluffy ice cream. It is best eaten immediately, but can be frozen once made.

Serves 4

4 large ripe bananas, coarsely sliced and frozen for at least 8 hours
¾ cup coconut milk, poured into a freezer bag or ice cube tray and frozen for at least 8 hours

To serve
Pinch of cinnamon
Coconut chips

1. Remove the banana and coconut milk from the freezer and place in a food processor. Blend for 3 to 4 minutes, until the mixture is completely smooth and has whipped to a light, ice cream consistency.

2. Serve immediately, sprinkled with a little cinnamon and some coconut chips.

If you choose to freeze the ice cream once it has been made, simply set aside at room temperature for 10 to 15 minutes before serving.

OIL

NUTTY GRANOLA

*VEGETARIAN *DAIRY-FREE

Making your own granola is not just cheaper, it is far more delicious than anything store-bought. Double the benefit by serving with coconut yogurt or coconut milk, or scattered over your favorite smoothie (try the ones on pages 14 and 20).

Makes 12 to 16 servings

5 cups rolled oats
¼ cup coconut oil, melted
¼ cup maple syrup
1 teaspoon vanilla extract
1½ cups mixed seeds
  (e.g., pumpkin, sunflower,
  and chia)
1 cup mixed nuts
  (e.g., cashews, almonds,
  and pecans), coarsely
  chopped
1 cup dried berries
  (e.g., cherries, cranberries,
  and blueberries)

1. Preheat the oven to 325°F. Place the oats in a large mixing bowl and drizzle over the oil, maple syrup, and vanilla. Add the seeds and nuts, and give the mixture a good stir.

2. Sprinkle the granola onto two lined baking sheets and cook for 15 to 20 minutes, until crisp and golden.

3. Allow the granola to cool before stirring in the berries. Transfer to an airtight container and keep for up to a month.

BEET & HORSERADISH SOUP

*VEGETARIAN

This warming, seasonal soup is perfect for a wintry lunch. The beets are full of antioxidants and bolstered by the vitamin C and zinc in the horseradish—great for keeping winter colds at bay. Either serve the horseradish swirled into the soup or in a separate bowl so everyone can have as much or as little as they like.

Serves 4

1 tablespoon coconut oil
1 onion, finely chopped
2¼ pounds beets, peeled and
   coarsely chopped
A few sprigs of thyme
1 quart good-quality
   vegetable stock
Sea salt and freshly ground
   black pepper

To serve
½ cup sour cream or yogurt
2 teaspoons grated
   horseradish
A handful of parsley,
   coarsely chopped

1. Heat the oil in a large saucepan over low to medium heat. Add the onion and cook, stirring occasionally, for 5 to 6 minutes, until softened. Stir in the beets and continue to cook for another 2 to 3 minutes. Add the thyme to the pan, pour in the stock, and simmer for 25 to 30 minutes, until the beets are tender.

2. In a food processer or blender, pulse the mixture until smooth and reheat slightly. Season to taste.

3. Divide the soup into individual bowls. Whisk the sour cream or yogurt with the horseradish until smooth and swirl into the soup, sprinkling with the parsley.

If you can't find it fresh, it's perfectly fine to use horseradish from a jar, but choose grated, hot horseradish over cream.

SWEET POTATO
& HARISSA SOUP
*VEGETARIAN *DAIRY-FREE

Sweet potatoes meld deliciously with rich, creamy coconut and earthy spices in this silky and satisfying soup. The soup also freezes brilliantly.

Serves 4

1 tablespoon coconut oil
1 onion, finely chopped
1 teaspoon ground cumin
2 pounds sweet potatoes, peeled and finely chopped
1 quart good-quality vegetable stock
1 cup coconut milk
Sea salt and freshly ground black pepper

To serve
2 teaspoons good-quality harissa paste
Pinch of red pepper flakes

1. Heat the oil in a large saucepan over low to medium heat. Add the onion and cook, stirring occasionally, for 5 to 6 minutes, until softened. Add the cumin and cook for another minute, until you can smell the spices. Add in the sweet potato and cook for a minute or so, until slightly golden.

2. Pour in the stock and coconut milk and simmer for 15 to 20 minutes, until the sweet potato is tender. Transfer the soup to a food processer and blend until smooth. Season to taste.

3. Serve in individual bowls with a swirl of harissa and the red pepper flakes.

Try pairing this soup with the Herbed Flatbreads on page 63.

BAKED MEXICAN EGGS

*VEGETARIAN *DAIRY-FREE *GLUTEN-FREE

These richly savory eggs make a deeply satisfying brunch. If you want to get ahead, simply prepare the tomato sauce the day before and reheat before dropping in the eggs.

Serves 2

2 teaspoons coconut oil
1 red onion, finely sliced
2 garlic cloves, crushed
1 red bell pepper, finely sliced
½ red chile, finely chopped
1 (14-ounce) can chopped
   tomatoes
Sea salt and freshly ground
   black pepper
1½ cups kale leaves
   (tough stems removed),
   finely chopped
4 large eggs

To serve
A handful of cilantro

1. Preheat the oven to 400°F. Heat the oil in a medium-sized, ovenproof frying pan over low to medium heat. Add the onion and cook for 6 to 8 minutes, until softened. If the onions start to stick and burn during cooking, add a splash of water.

2. Add the garlic, peppers, and chile to the pan and cook for 2 to 3 minutes, until the garlic is fragrant and the peppers start to take on a little color. Pour in the chopped tomatoes and simmer for 5 minutes, until the sauce has reduced slightly and the peppers are tender. Season lightly.

3. Stir in the kale and make four wells in the sauce. Crack an egg into each well and transfer to the oven for 4 to 5 minutes, or until the eggs are cooked to your liking. To serve, sprinkle with the cilantro, put it on the table, and dive in.

MUSHROOM, FETA & SPINACH OMELET

*VEGETARIAN *GLUTEN-FREE

This omelet is wonderfully simple to prepare and the protein in the eggs and feta will keep you satisfied for hours. Eggs and mushrooms are also high in selenium and vitamin D.

Makes 2

3 teaspoons coconut oil
1 garlic clove, finely sliced
1¼ cups brown mushrooms, sliced
⅓ cup feta cheese, crumbled
4 large eggs, beaten
Sea salt and freshly ground black pepper

To serve
A large handful of baby spinach leaves
Lemon juice

1. Heat a teaspoon of oil in a medium, nonstick frying pan over medium heat. Add the garlic and cook for a few minutes. Then add the mushrooms and cook for another 2 to 3 minutes, turning occasionally, until golden. Remove from the pan with a slotted spoon and transfer to a bowl. Mix the mushrooms with the feta cheese and set aside.

2. Return the pan to the heat and add another teaspoon of the oil. Reduce the heat slightly, season the beaten egg with a pinch of salt and pepper, and pour in half of the beaten egg. Swirl the egg around the pan and mix with a spatula, until the omelet begins to set. Once the egg is almost set (this should take about two minutes), sprinkle half the mushroom and feta mix onto one side of the omelet and top with the spinach and a squeeze of lemon. Carefully fold the other half over and slide onto a plate.

3. Repeat with the remaining egg to make a second omelet and serve on individual plates with the spinach and a squeeze of lemon juice on the side.

PAPRIKA & LIME SWEET POTATO WEDGES

*VEGETARIAN *DAIRY-FREE *GLUTEN-FREE

A light and smoky take on traditional fries, these wedges are delicious simply served with a spicy mayonnaise or as a side dish, plus they're packed with vitamin A for healthy skin. They pair really well with the Turkey Burgers on page 57.

Serves 4

1¼ pounds sweet potatoes, scrubbed and cut into eight wedges per potato
Zest of 2 limes
Sea salt and freshly ground black pepper
1½ teaspoons sweet smoked paprika
1½ tablespoons coconut oil, melted
¼ cup good-quality mayonnaise
Tabasco sauce

1. Preheat the oven to 400°F. Place the wedges in a large bowl and scatter over the zest of one lime, a pinch of salt and pepper, and the paprika. Drizzle with the oil and toss the wedges until well-coated.

2. Lay the wedges out in a single layer on a baking sheet and cook for 20 to 25 minutes, turning halfway through until crisp and golden.

3. Mix the mayonnaise with the remaining lime zest and a few drops of Tabasco sauce and serve with the potato wedges.

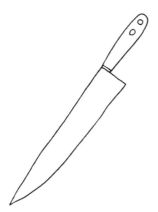

SLOW-ROASTED TOMATO SAUCE

*VEGETARIAN *DAIRY-FREE *GLUTEN-FREE

Roasting tomatoes slowly brings out their natural sweetness beautifully. This rustic sauce is delicious as it is, paired with any pasta of your choice, or blended to make a finer sauce.

Serves 6 to 8

2¼ pounds ripe or overripe tomatoes, halved
1 small onion, chopped
4 fat garlic cloves, finely chopped
2 tablespoons coconut oil, melted
Pinch of sugar
Sea salt and freshly ground black pepper

To serve
A handful of basil leaves

1. Preheat the oven to 350°F. Lay the tomatoes, cut-side up, on one or two baking sheets in a single layer. Scatter the onions over the top and dot each tomato with a little of the chopped garlic. Drizzle over the oil and season with the sugar, salt, and pepper.

2. Roast in the oven for 50 to 60 minutes, until the tomatoes are completely soft and the skins slightly burnished.

3. The tomatoes can be left whole and stirred through a pasta of your choice, with the basil leaves scattered over, or blended to make a chunky sauce.

The roasted tomatoes can be kept halved or blitzed for up to a week in the fridge, covered.

PEARL BARLEY & SQUASH RISOTTO

*VEGETARIAN *DAIRY-FREE

Pearl barley offers a wholesome and nutritious alternative to rice in this delicate risotto. It has a far higher percentage of fiber than brown and white rice, which aids good digestion.

Serves 4

3 teaspoons coconut oil
1 small butternut squash, (1½ to 2 pounds), peeled and cut into ½-inch cubes
2 shallots, peeled and finely chopped
2 garlic cloves, crushed
A small handful of sage leaves, coarsely chopped (reserve a little for garnish)
1½ cups pearl barley
6 cups good-quality vegetable stock
Juice of ½ lemon

To serve

A handful of parsley, finely chopped
Sea salt and freshly ground black pepper

1. Heat 2 teaspoons of the oil in a large frying pan over medium heat. Add the squash and cook for 3 to 4 minutes, turning from time to time, until golden. Remove the squash from the pan and set aside.

2. Add the remaining oil to the pan, reduce the heat, and stir in the shallots. Cook for 4 to 5 minutes, until softened. Add the garlic and half the sage and continue to cook for another minute, until the garlic is fragrant.

3. Pour the pearl barley into the pan and stir for 1 to 2 minutes to toast the barley slightly. Add a ladle of stock, stirring occasionally, until the liquid is absorbed. Continue to add the stock like this and keep stirring, then add the lemon juice.

4. Stir the squash back into the risotto after 15 minutes and cook for another 15 minutes or so, until the barley is tender and you have used up all the stock.

5. Divide the risotto between individual plates, top with the remaining sage, parsley, and a pinch of salt and pepper.

SHRIMP WITH ASIAN GREENS & NOODLES

*DAIRY-FREE *GLUTEN-FREE

This flavorful, nutritious stir-fry can be thrown together in less than 30 minutes—ideal for a light midweek dinner.

Serves 4

7 ounces brown rice noodles
1 tablespoon coconut oil
½-inch piece of fresh ginger, peeled and finely grated
2 garlic cloves, crushed
2 bok choy, stems and leaves coarsely sliced
1½ cups sugar snap peas, sliced in half
6 ounces raw jumbo shrimp, deveined if necessary

For the sauce
2 tablespoons tamari soy sauce
2 teaspoons coconut or palm sugar
Juice of 1 lime
1 red chile, finely chopped

To serve
2 scallions, finely chopped
1 tablespoon toasted sesame seeds

1. Start by soaking the noodles in a bowl of boiling water for 5 minutes, until almost tender. Drain, rinse under cold water, and set aside.

2. Now make the sauce: Stir together the tamari, sugar, lime juice, and chile and set aside.

3. Heat the oil in a large frying pan or wok over medium heat. Throw in the ginger and garlic, stir-fry for a minute or so, and then stir in the bok choy stems and sugar snap peas. Stir-fry for a minute, then add the shrimp, cooking for another minute before pouring in the sauce. Allow the sauce to bubble for a minute or so, until the shrimp is opaque and the sauce has reduced slightly.

4. Add the bok choy leaves and stir for a minute until they start to wilt slightly. Toss the noodles in the pan to reheat, mixing everything together, then divide between individual plates, scattered with the scallions and sesame seeds.

MISO SALMON & EGGPLANT SKEWERS

*DAIRY-FREE

These skewers are so simple to prepare and yet incredibly delicious. The white miso caramelizes under the broiler to a perfectly balanced savory sweetness that the whole family will love.

Serves 4

For the marinade

3 tablespoons white miso paste
2 teaspoons coconut oil, melted
2 tablespoons mirin
1 tablespoon tamari soy sauce
1-inch piece of fresh ginger, peeled and finely grated

For the skewers

1 eggplant, cut in half lengthwise, each half cut into 8 chunks
4 salmon fillets, skinned and cut in half horizontally

To serve

1 tablespoon sesame seeds
2 scallions, finely sliced
A handful of herbs (cilantro works well)

1. First make the marinade: Spoon the miso into a bowl and whisk in the coconut oil, mirin, tamari, and ginger.

2. Next add the eggplant and salmon to the bowl and stir well, until they are well coated in the marinade. Set aside in the fridge for up to an hour or cook immediately.

3. Preheat the broiler to high. Thread one chunk of eggplant, followed by a piece of salmon onto a skewer. Finish with a final chunk of eggplant before transferring to a foil-lined baking sheet.

4. Broil the skewers for 1 to 2 minutes on each side, until the fish is cooked through and everything is nicely caramelized.

5. Divide between four plates, scattered with the sesame seeds, scallions, and herbs.

VEGAN CHILI WITH GUACAMOLE & SALSA

*VEGETARIAN *DAIRY-FREE *GLUTEN-FREE

Vegan chili can be just as delicious and nutritious as its meaty counterpart—rich, earthy spices, delicate sweet potatoes, and creamy beans, finished with spicy guacamole and salsa, create a wonderfully balanced meal.

Serves 4 to 6

For the chili

3 teaspoons coconut oil
1¼ pounds sweet potatoes, scrubbed and cut into ¾-inch pieces
1 teaspoon ground cumin
1 teaspoon mild chile powder
1 teaspoon smoked paprika
1 onion, finely chopped
2 garlic cloves, crushed
1 green bell pepper, coarsely chopped
2 (14-ounce) cans chopped tomatoes
1 tablespoon tomato paste
1 (14-ounce) can kidney beans, drained and rinsed
1 (14-ounce) can black beans, drained and rinsed
Sea salt and freshly ground black pepper

cont. on next page

1. First make the chili: Heat 1 teaspoon of the oil in a large frying pan over medium heat. Stir in the sweet potato chunks and cook for 4 to 5 minutes, until nicely golden on all sides. Remove from the pan with a slotted spoon and set aside.

2. Reduce the heat slightly and add the remaining oil. Once melted, add the cumin, chile powder, and smoked paprika. Cook the spices for a minute or so, until lightly toasted and aromatic.

3. Reduce the heat again, add the onion, and cook very gently, stirring from time to time, for 5 to 6 minutes, until softened.

Stir in the garlic and green bell pepper and continue to cook for another 2 to 3 minutes, until the pepper has taken on a little color and the garlic has softened.

4. Add the chopped tomatoes and tomato paste, transfer the sweet potato back to the pan, turn the heat up a little, and simmer for 6 to 8 minutes, until the mixture has slightly reduced and thickened. Stir in the drained beans and simmer for another 5 minutes, seasoning to taste. By now, the mixture should be fairly thick and the sweet potato tender.

cont. on next page

For the guacamole
1 large, ripe avocado
½ shallot, finely chopped
½ red chile, finely diced
Juice of ½ lime (reserve half
   for the salsa)
1 to 2 tablespoons coarsely
   chopped cilantro leaves

For the salsa
1 cup cherry tomatoes,
   quartered
½ shallot, finely chopped
½ red chile, finely diced
Lime juice
1 to 2 tablespoons coarsely
   chopped cilantro leaves

5. Now make the guacamole: Scoop the avocado flesh into a bowl and add half the shallot, half the chile, and half the lime juice. Mash the mixture with a fork and season with a little salt.

6. For the salsa, place the tomatoes in a separate bowl and add the remaining shallot, chile, and lime juice. Season to taste with salt and stir to combine.

7. Divide the chili into individual bowls and serve with a dollop of the guacamole and salsa.

GRILLED CHICKEN WITH GREEK SALAD
*GLUTEN-FREE

This Greek-inspired dish is reminiscent of warm summer evenings. While delicious served warm, the chicken can be cooked in advance and eaten cold with the salad. The onion, vinegar and tomatoes add an antioxidant boost.

Serves 4

4 medium chicken breasts, skinned
Juice and zest of 1 lemon
A small handful of oregano, finely chopped (if using dried oregano, ½ teaspoon will do)
Sea salt and freshly ground black pepper
2 teaspoons coconut oil, melted

For the salad
1 red onion, finely sliced
1 tablespoon white wine or cider vinegar
7 ounces ripe vine tomatoes
⅓ cup black kalamata olives, pitted
1 cucumber, peeled, seeded, and coarsely sliced
½ cup good-quality feta cheese, crumbled
2 teaspoons fresh oregano, coarsely chopped, or 1 teaspoon dried

1. Lay the chicken breasts between two sheets of parchment paper and beat with a rolling pin to flatten the meat out slightly.

2. Place the chicken in a bowl and pour over the lemon juice and zest and oregano and season with salt and pepper. Marinate the chicken for at least an hour, or preferably overnight in the fridge.

3. When ready to cook, remove the chicken from the fridge 15 minutes before cooking. Heat a grill pan over medium heat. Drizzle the coconut oil over the chicken, and once the pan is fairly hot, cook the chicken for 4 to 5 minutes on each side, until the chicken juices run clear.

4. Now make the salad: Place the onion in a salad bowl and pour in the vinegar. Add the tomatoes, olives, cucumber, and feta and stir.

5. Divide the chicken between four plates, sprinkle with the oregano, and serve alongside the salad.

TURKEY BURGERS & FENNEL SLAW
*GLUTEN-FREE

These quick and flavorful burgers are great for a quick midweek supper or a summer barbecue. The crisp and citrusy salad makes a delicious alternative to creamy coleslaw.

Serves 4

For the burgers
1¼ pounds ground turkey
3 scallions, finely sliced
⅓ cup sharp cheese (e.g. feta or Cheddar), crumbled
2 tablespoons parsley, finely chopped
Sea salt and freshly ground black pepper
2 teaspoons coconut oil

For the slaw
1 large fennel bulb, finely sliced
1 watermelon radish, finely sliced
1 shallot, finely sliced
¼ red cabbage, finely sliced
1 orange, peeled and segmented
½ teaspoon fennel seeds
2 teaspoons olive oil
3 tablespoons pecans, coarsely chopped

1. First make the burgers: Combine the ground turkey with the scallions, cheese, and parsley, then season with salt and pepper. Shape the mixture into four patties and set aside.

2. Heat the oil in a large, nonstick frying pan over medium heat. Add the burgers and cook for 5 to 6 minutes on each side, until golden and completely cooked through.

3. To make the slaw, mix all the ingredients together and season with a little salt and pepper.

4. Divide the burgers between four plates and serve with the slaw.

Choose ground turkey as it is leaner than chicken.

FLOUR

SEEDED COCONUT FLOUR BREAD

*VEGETARIAN *GLUTEN-FREE *DAIRY-FREE

This richly seeded bread doesn't require any kneading or proving—just mix and bake. It's also a great gluten- and dairy-free staple. It's wonderful either simply sliced or toasted with nut butter.

Makes a 2lb loaf

1⅔ cups ground almonds
1 cup coconut flour
1½ teaspoon baking powder
½ teaspoon baking soda
Sea salt
½ cup mixed seeds
    (e.g. poppy seeds, sunflower
    seeds, and flax seeds)
6 large eggs
¼ cup coconut oil, melted
1½ tablespoons lemon juice

1. Preheat the oven to 375°F. Place the almonds in a large mixing bowl, breaking up any lumps as you go. Sift over the coconut flour, baking powder, and baking soda, stir in the salt and seeds, and make a well in the center.

2. Beat in the remaining ingredients and transfer to a 2-pound loaf pan lined with parchment paper. Bake for 0 to 45 minutes, until golden and cooked through.

3. Allow the bread to cool in the pan for 10 minutes before transferring to a wire rack to cool completely. The bread will keep in an airtight container for up to three days.

STICKY DATE LOAF

*VEGETARIAN *GLUTEN-FREE *DAIRY-FREE

A healthier take on traditional sticky toffee pudding, this loaf is equally delicious and full of fiber and protein to boot. Enjoy warm with a generous helping of "toffee" sauce.

Serves 12

1¼ cups medjool dates, pitted and finely chopped
1 cup boiling water
1 teaspoon baking soda
¼ cup coconut oil, melted
¼ cup coconut sugar
3 extra-large eggs, lightly beaten
1 teaspoon vanilla extract
1 cup gluten-free all-purpose flour
¾ cup coconut flour
1 teaspoon baking powder
1 teaspoon xanthum gum

For the sauce
⅔ cup coconut cream
2 tablespoons coconut sugar
1 teaspoon vanilla extract

1. Preheat the oven to 350°F. Lightly grease a 2-pound loaf pan and line with parchment paper.

2. Place the chopped dates in a bowl, pour in the boiling water, and stir in the baking soda. Set aside until needed.

3. In a separate, large mixing bowl, beat the oil, sugar, eggs, and vanilla together until smooth. Sift over the flours, baking powder, and xanthum gum and fold through until you have a thick batter. Add the dates, along with their soaking water, and fold through to loosen the mixture.

4. Pour into the prepared pan and bake for 45 to 50 minutes, until risen and golden. A skewer inserted into the center should come out clean.

5. Remove the cake from the oven and let it cool in the pan for 10 minutes before turning out.

6. Next make the sauce: Pour the coconut cream and sugar into a saucepan, place over medium heat, and simmer for 5 to 6 minutes until slightly thickened. Stir through the vanilla and serve drizzled over a slice of the loaf. The cake is best served warm but will keep in an airtight container for up to three days.

Check the cake after 30 minutes—if it's looking a little dark, cover with a piece of foil and return to the oven.

HERBED FLATBREADS

*VEGETARIAN *GLUTEN-FREE *DAIRY-FREE

These simple flatbreads are so easy to make and a perfect vehicle for dips, dahls, and soups. Try them with the Sweet Potato & Harissa Soup on page 39.

Makes 4

½ cup coconut flour
½ teaspoon xantham gum
1 teaspoon baking powder
Sea salt
3 large eggs, beaten
2 tablespoons mixed herbs
  (e.g. parsley, thyme,
  rosemary, and sage),
  finely chopped
2 tablespoons melted
  coconut oil

1. Sift together the flour, xantham gum, baking powder, and a good pinch of salt into a bowl.

2. Make a well in the center and add the egg, herbs, and coconut oil. Gradually beat in the wet ingredients to create a thick batter.

3. Heat a nonstick frying pan over medium heat. With damp hands, shape a quarter of the mixture into a flatbread or pita shape. Place in the pan and dry-fry for 2 to 3 minutes per side or until golden brown. Repeat until you have made four flatbreads.

SMOKY CORN FRITTERS

*VEGETARIAN *GLUTEN-FREE *DAIRY-FREE

Smoky, sweet, and nutritious fritters are a wonderful brunch-time treat, full of vitamins, minerals and antioxidants from the avocado and tomato.

Serves 4

For the fritters

1 (14-ounce) can corn, or 2½ cups cut from cooked cobs
⅓ cup coconut flour
3 tablespoons gluten-free flour
1½ teaspoons baking powder
¼ teaspoon sea salt
1 teaspoon smoked paprika
2 scallions, finely chopped
1 red chile, finely chopped
3 large eggs, lightly beaten
½ to ⅔ cup almond milk
2 teaspoons coconut oil

For the salsa

½ cup tomatoes, coarsely chopped
1 shallot, finely chopped
A handful of parsley and cilantro, coarsely chopped
Juice of 1 lime
1 large ripe avocado, pit removed, peeled, and finely sliced

1. Place 1¾ cups of the corn in a mixing bowl and sift over the flours, baking powder, salt, and paprika. Add the scallions and chile to the bowl and stir everything to combine.

2. Pour the remaining corn into a food processor along with the eggs and pulse until the mixture is fairly smooth. Pour this over the corn and flour and stir until the mixture forms a batter. Add the milk, a little at a time, to achieve a fairly thick, pourable consistency.

3. Heat the oil in a nonstick frying pan over medium heat and drop 2 heaping tablespoons of the batter into the pan. Cook for 2 to 3 minutes on each side, until golden and cooked through. Repeat with the remaining batter. Set aside and keep warm.

4. To make the salsa, mix the tomatoes, shallot, herbs, and lime together and pour over the avocado.

5. Serve a stack of fritters on a big plate, topped with the salsa, and dive in.

PEANUT BUTTER
COOKIES *VEGETARIAN *GLUTEN-FREE

These nutty cookies are a perfect afternoon treat. They
tend to rise quite a bit during cooking, so patting them
into discs will help to retain a cookie-like shape.

Makes 16 to 20

¾ cup ground almonds
½ cup coconut flour
1 teaspoon baking powder
1 teaspoon xanthum gum
½ cup crunchy peanut butter
⅓ cup coconut oil, melted
3 large eggs, beaten
⅓ cup coconut sugar
¼ cup maple syrup
1 teaspoon vanilla extract
2 ounces good-quality
   dark chocolate, minimum
   70 percent cocoa solids,
   finely chopped

1. Preheat the oven to 350°F.
Line two baking sheets with
parchment paper.

2. Sprinkle the ground almonds
into a large mixing bowl,
breaking up any lumps as you go.
Sift over the coconut flour, baking
powder, and xanthum gum. Now
make a well in the center.

3. Place the peanut butter into the
well, then add the oil, eggs, sugar,
maple syrup, and vanilla. Beat
the wet ingredients into the dry
to create a thick batter. Stir in the
chocolate.

4. Pat tablespoons of the
mixture into discs using your
palms before transferring to the
prepared pans, leaving a gap of at
least ¾-inch between each disc.

5. Bake the cookies for
10 to 12 minutes, until risen
and golden. Allow the cookies to
cool on the pans for 10 minutes
before transferring to a cooling
rack. The cookies are delicious
warm but will keep in an airtight
container for up to three days.

PANCAKES WITH MAPLE BANANAS

*VEGETARIAN *GLUTEN-FREE *DAIRY-FREE

Once you've tried these fluffy pancakes with rich, caramelized bananas, they are sure to become a breakfast or brunch staple. Coconut flour produces a far lighter pancake than its heavy, glutinous counterpart.

Serves 2

½ cup coconut flour
1 teaspoon baking powder
Sea salt
3 tablespoons coconut oil, melted, plus a little extra for frying
4 extra-large eggs, lightly beaten
½ cup almond or coconut milk
2 bananas, sliced
6 tablespoons maple syrup

1. First make the batter: Sift the flour and baking powder into a mixing bowl. Stir in the salt and make a well in the center of the flour. Pour in the oil and eggs and whisk to gradually incorporate the flour. Whisk in enough milk to achieve a heavy cream consistency.

2. Heat a little coconut oil in a large, nonstick frying pan and drop 2 tablespoons of the batter into rounds, leaving a little space between each. Cook the pancakes for 1 to 2 minutes. When the edges are set and there are bubbles covering the surface of the pancakes, flip them over and continue to cook for another 1 to 2 minutes until risen, golden,

and fluffy. Set aside and keep warm while you repeat the process until the batter has been used up.

3. Keep the pan on the heat and add a little more oil. Add the sliced bananas and cook for 30 seconds, until they start to take on a little color. Pour in 2 tablespoons of the maple syrup, turn the bananas over, and cook for another 30 to 60 seconds, until golden and syrupy.

4. Divide the pancakes between two plates and serve with the bananas and remaining maple syrup.

ORANGE & POPPY SEED MUFFINS

*VEGETARIAN *DAIRY-FREE

These fiber- and protein-rich muffins are an excellent breakfast or afternoon snack. When in season, try using sweet, ruby-fleshed blood oranges for variety.

Makes 12

2 cups spelt flour
½ cup coconut flour
1 teaspoon baking powder
½ teaspoon baking soda
4 large eggs, lightly beaten
⅓ cup coconut oil
    or 5 tablespoons butter,
    melted
½ cup coconut sugar
2 oranges (blood oranges,
    if in season), finely zested,
    flesh peeled and cut into
    six slices each
2 tablespoons poppy seeds
½ to ¾ cup almond or
    coconut milk

1. Preheat the oven to 350°F. Line a 12-hole muffin pan with paper liners.

2. Sift the flours, baking powder, and baking soda into a large bowl. Make a hole in the center of the dry ingredients and pour in the eggs, coconut oil or butter, sugar, orange zest, and poppy seeds. Use a metal spoon to fold the wet ingredients into the dry, adding in the milk to loosen to a dropping consistency.

3. Fill each muffin liner two thirds full and top with a slice of orange. Bake for 18 to 20 minutes, until a skewer inserted into the center of each muffin comes out clean.

4. Serve warm or cold. These muffins will keep in an airtight container for up to two days.

Using blood oranges gives you an extra boost of vitamin C— they can contain up to 40 percent more than other oranges.

PISTACHIO BROWNIES

*VEGETARIAN *GLUTEN-FREE

Fudgy chocolate brownies are the last word in indulgence. Try a less sinful treat with this recipe, packed with good fats, high-fiber coconut flour, and antioxidant-rich dark chocolate.

Makes 16 squares

¾ cup coconut oil or 12 tablespoons (1 stick + 4 tablespoons) butter
7 ounces good-quality dark chocolate, minimum 70 percent cocoa solids, coarsely chopped
½ cup ground almonds
¾ cup coconut flour
2 tablespoons cocoa powder
¾ cup coconut sugar
¼ cup maple syrup
3 large eggs, lightly beaten
1 teaspoon vanilla extract
¾ cup pistachios, coarsely chopped

1. Preheat the oven to 350°F. Grease and line an 8-inch square pan with parchment paper.

2. Add the oil or butter and chocolate to a pan. Place over very low heat and allow the mixture to melt, stirring occasionally. As soon as the chocolate and fat has melted, remove from the heat and set aside.

3. Sprinkle the ground almonds into a large mixing bowl, breaking up any lumps as you go. Sift over the coconut flour and cocoa powder.

4. In a separate bowl, beat together the coconut sugar, maple syrup, and eggs with a handheld mixer until foamy and paler in color—this will take about 3 to 4 minutes.

5. Pour the egg mixture over the dry ingredients and use a large metal spoon to gently fold together. Pour in the melted chocolate, vanilla extract, and three quarters of the pistachios and carefully fold through, so as not to knock out too much air.

6. Place the mixture into the prepared pan, sprinkle with the remaining pistachios, and bake for 20 to 25 minutes, until the brownies are crisp on top and just set. Allow the brownies to cool slightly before cutting into squares. They will keep in an airtight container for up to three days.

DRIED

GRILLED
VEGETABLES

*VEGETARIAN *GLUTEN-FREE *DAIRY-FREE

Great as a side dish or on their own, these grilled vegetables are paired with a sour kick from the sumac and lemon and tempered by the nutty sweetness of coconut.

Serves 4

For the marinade
2 teaspoons coconut oil, melted
2 garlic cloves, crushed
¼ teaspoon sumac
Zest of 1 lemon (juice of 1 lemon reserved to serve)
Sea salt and freshly ground black pepper

For the vegetables
1 fennel bulb, cut into eight wedges
7 ounces (about 1 bunch) asparagus, woody ends discarded
7 ounces broccolini, ends trimmed
1½ cups cherry tomatoes, on the vine
⅓ cup coconut chips

1. First make the marinade: Pour the coconut oil into a large mixing bowl and add the garlic, sumac, and lemon zest. Season with a pinch of salt and pepper.

2. Add the fennel, asparagus, broccolini, and tomatoes to the bowl and toss in the marinade to coat.

3. Place a grill pan over medium heat and, when hot, lay some of the vegetables on the pan in a single layer. Cook for about 2 minutes on each side, until lightly burnished and tender. Set aside and repeat with the remaining vegetables.

4. Sprinkle the coconut chips over the pan and toast briefly until golden.

5. Divide the vegetables between individual plates and drizzle with the lemon juice and toasted coconut.

BROILED COD WITH RADISH & HERBS

*GLUTEN-FREE *DAIRY-FREE

This delicate and vibrant supper requires a close eye on time but the results are fabulous. The radish salad and coconut chips add a good source of fiber.

Serves 4

4 skinless cod loin fillets, weighing about 6 ounces each
2 tablespoons olive or coconut oil
Juice and zest of 1 lemon
1 shallot, finely chopped
1½ cups radishes, halved
Sea salt and freshly ground black pepper
⅓ cup coconut chips

To serve
A handful of mixed herbs (e.g. basil, dill, and parsley), coarsely chopped
3½ ounces salad leaves

1. Preheat the broiler to medium. Brush the fish fillets with half of the oil and place on a baking sheet lined with foil. Sprinkle with the lemon zest and broil for 10 to 12 minutes, until the flesh just flakes to the touch. Set aside and keep warm.

2. Meanwhile, pour the lemon juice over the shallots and set aside. Toss the radishes in the remaining oil, season with a little salt and pepper, and broil for 3 to 4 minutes, until lightly colored. Sprinkle the coconut chips over the radishes for the final 30 seconds of cooking.

3. Divide the fish between four plates, scattered with the shallots, radishes, and coconut. Serve the herbs and salad alongside.

SPICY CHICKEN & CUCUMBER SALAD

*GLUTEN-FREE *DAIRY-FREE

The light and nutty crunch of shredded coconut offers a delicious and high-fiber alternative to breaded chicken. Serve with a spicy, cooling cucumber salad for a light supper.

Serves 4

1 large egg, lightly beaten
⅔ cup shredded coconut
¼ teaspoon red pepper flakes
Zest of 1 lime, wedges reserved for garnish
Sea salt and freshly ground black pepper
4 medium skinless chicken breasts
1 tablespoon coconut oil, melted

For the salad
1 tablespoon rice wine vinegar
1 teaspoon coconut or palm sugar
1 large cucumber, halved lengthwise, seeded and coarsely sliced, or 9 ounces baby cucumbers, sliced
A small handful each of cilantro, mint, and dill, coarsely chopped

1. Preheat the oven to 400°F. Prepare the chicken coating by filling a shallow dish with the beaten egg and another with the coconut and lime zest. Add a pinch of red pepper flakes to the coconut mix and season lightly with salt and pepper. Slice the lime into wedges and set aside.

2. Dip each chicken breast in the egg before rolling in the coconut mix to cover. Transfer to a baking sheet, drizzle with the coconut oil, and cook for 18 to 20 minutes, or until the crust is golden and chicken juices run completely clear.

3. Meanwhile, prepare the salad: Mix the vinegar with the sugar until completely dissolved. Toss the cucumber in the dressing and scatter with the herbs and remaining pepper flakes.

4. Divide the chicken and salad between individual plates alongside a wedge of lime.

PINEAPPLE CARPACCIO

*VEGETARIAN *GLUTEN-FREE *DAIRY-FREE

A carpaccio elevates a simple sliced fruit into something you can show off to guests. The crisp, toasted coconut and sugar create wonderful textures and flavors.

Serves 4 to 5

1 pineapple
⅓ cup coconut chips
Zest of 1 lime
1 tablespoon coconut sugar

1. To prepare the pineapple, slice off the base and leaves and stand upright on a cutting board. Slice off the skin downward in strips and remove any eyes with a small knife. Lay the pineapple on its side and slice thinly.

2. Toast the coconut chips in a small, nonstick pan until golden. Remove from the heat and set aside.

3. Lay the pineapple slices on a serving dish and sprinkle with the coconut. Mix the lime zest and sugar together, drizzle over the pineapple, and serve immediately.

DATE & COCOA
ENERGY BALLS

*VEGETARIAN *GLUTEN-FREE *DAIRY-FREE

These delicate energy balls are a lot like chocolate truffles, although infinitely better for you. Full of nutrients and good fats, enjoy before a workout or as an afternoon snack.

Makes 16 to 18

1¼ cups medjool dates, pitted
3 tablespoons cocoa powder
   (check it's dairy-free
   if needed)
1 tablespoon coconut oil
¼ teaspoon cinnamon
⅔ cup shredded coconut, plus
   2 tablespoons to serve

1. Place the dates, oil, cinnamon, coconut, and 1 tablespoon of the cocoa powder in a food processor and blend until fairly smooth.

2. Using damp hands, roll the mixture into individual balls. Sprinkle a plate with the remaining coconut and another with the cocoa powder. Roll half the balls in the coconut and half in the cocoa. Refrigerate to set.

These energy balls will keep in the fridge for up to 10 days. They are perfect for snacking when the cravings hit!

OAT & SOUR CHERRY COOKIES *VEGETARIAN *DAIRY-FREE

Soft and chewy oat cookies jeweled with plump, sour cherries go wonderfully with a cup of coffee or hot chocolate (try the Cardamom Hot Chocolate on page 17). They're also ideal for lunchboxes.

Makes 12 to 16 cookies

⅓ cup coconut flour
½ teaspoon baking powder
½ teaspoon ground cinnamon
½ cup porridge oats
¼ cup shredded coconut
⅓ cup coconut oil or
  5 tablespoons butter, melted
⅓ cup coconut or light brown sugar
¼ cup maple syrup
1 extra-large egg, lightly beaten
½ cup dried sour cherries, coarsely chopped

1. Preheat the oven to 350°F. Line two large baking sheets with parchment paper and set aside.

2. Sift the coconut flour, baking powder, and cinnamon into a large mixing bowl. Stir in the porridge oats and shredded coconut and make a well in the center. Pour in the melted oil or butter, followed by the sugar, maple syrup, and egg. Beat the wet ingredients into the dry to make a thick, sticky dough. Stir in the cherries.

3. Transfer heaping tablespoons of the mixture onto the baking sheets, leaving a space of at least 1½-inches between each one.

4. Bake in the oven for 12 to 14 minutes, until they are golden and risen. Let cool on the pans for 10 minutes, before transferring to a cooling rack. The cookies will keep in an airtight container for up to three days.

CRANBERRY & PUMPKIN SEED BARS

These oaty bars make an ideal, energy-boosting snack. Rustle up a batch and keep on hand for that mid-afternoon slump.

Makes 12

6 medjool dates, pitted
¼ cup hot water
¼ cup coconut oil, melted
2 tablespoons crunchy peanut butter
2 tablespoons honey
1 large banana, coarsely chopped
2 cups rolled oats
¾ cup shredded coconut
½ cup dried cranberries
½ cup pumpkin seeds

1. Preheat the oven to 325°F. Grease and line an 8-inch square pan with parchment paper.

2. Soak the dates in the hot water for 10 minutes, to soften. Place them in a food processor, along with their soaking water, then add the coconut oil, peanut butter, honey, and banana. Blend until fairly smooth.

3. Pour the mixture into a bowl and stir in the oats, coconut, cranberries, and pumpkin seeds. Spoon into the prepared pan and bake for 50 to 60 minutes, until golden brown and slightly crisp. Let cool before cutting into 12 pieces. The bars will keep in an airtight container for up to four days.

For extra depth of flavor, lightly toast the coconut and oats in a dry frying pan before adding to the bar mix.

HEALTHY, SALTED BANOFFEE TREATS

*VEGETARIAN *GLUTEN-FREE

These desserts take a little bit of effort, but they are well worth it. A healthier version of this classic dessert is a revelation—perfect for entertaining.

Serves 4

For the "toffee"
¾ cup coconut cream
¼ cup coconut sugar
Pinch of sea salt

For the base
⅓ cup shredded coconut
⅓ cup pecans
1 medjool date, pitted
1 teaspoon coconut oil

For the topping
½ cup coconut cream, chilled
2 bananas, sliced
¾ ounce good-quality dark chocolate, minimum 70 percent cocoa solids, grated

1. First make the toffee: Pour the coconut cream and sugar into a small saucepan and place over medium heat. Bring up to a boil and simmer for 6 to 8 minutes, stirring frequently, until thickened. Stir in the salt and set aside to cool.

2. When the toffee has cooled to room temperature, transfer it to a bowl, cover with plastic wrap, and refrigerate for at least two hours. It can be made up to a week in advance.

3. For the base, place all the ingredients in a food processor and pulse until fairly smooth and sticky. Sprinkle the mixture into four glasses or ramekins and let set in the fridge for an hour.

4. For the topping, whip the coconut cream with a handheld mixer until thickened. Scoop a spoonful of the toffee onto each base and spread it out a little to reach the edges. Scatter with the banana slices, dollop some coconut cream on each one, and sprinkle with the chocolate. Serve immediately.

INDEX

A

almond & coconut green
smoothie 14–15

B

baby corn & lime with chicken
soup 24–5
banana: banana & berry smoothie
bowl 20–1
  healthy, salted banoffee treats 90–1
  pancakes with maple bananas 68–9
beet & horseradish soup 37
benefits of coconut 9
berries: banana & berry smoothie
bowl 20–1
  coconut & berry Bircher
  muesli 18–19
bread: herbed flatbreads 63
  seeded coconut flour bread 60–1
brownies, pistachio 72–3
burgers, turkey 56–7
butternut squash & pearl barley
risotto 46–7

C

cardamom hot chocolate 16–17
carpaccio, pineapple 82–3
cauliflower & garlic soup 28
cherry: oat & sour cherry
cookies 86–7
chia, peach & vanilla puddings 29
chicken: chicken soup with baby
corn & lime 24–5
  grilled chicken with Greek
  salad 55
  spicy chicken & cucumber salad 80–1
chili: vegan chili with guacamole &
salsa 52–4

chocolate: cardamom hot
chocolate 16–17
cocoa & date energy balls 84–5
coconut chips 76, 79, 82
coconut cream 10, 62, 91: coconut
creams & grilled pineapple 30–1
coconut flour 7, 10: herbed
flatbreads 63
  oat & sour cherry cookies 87
  orange & poppy seed muffins 70
  pancakes with maple bananas 69
  peanut butter cookies 67
  pistachio brownies 73
  seeded coconut flour bread 60–1
  smoky corn fritters 64–5
  sticky date loaf 62
coconut milk 7, 10: banana & berry
smoothie bowl 20
  cardamom hot chocolate 17
  cauliflower & garlic soup 28
  chicken soup with baby
  corn & lime 24
  coconut & berry Bircher muesli 18
  coconut creams & grilled
  pineapple 30
  creamy red lentil & coconut dahl 27
  instant ice cream 32
  orange & poppy seed muffins 70
  pancakes with maple bananas 69
  peach & vanilla chia puddings 29
  sweet potato & harissa soup 39
  shrimp tacos with coconut & lime 23
coconut oil 11, 23–4, 27–8, 36–7, 39,
41–2, 44–5, 47, 49–50, 53, 55, 62–4, 67,
70, 73, 76, 79–80, 87–8
coconut sugar 10: chicken soup with
baby corn & lime 24

coconut creams & grilled
pineapple 30
healthy, salted banoffee treats 90–1
oat & sour cherry cookies 87
orange & poppy seed muffins 70
peanut butter cookies 67
pineapple carpaccio 82
pistachio brownies 73
shrimp with Asian greens &
noodles 48–9
spicy chicken & cucumber salad 80–1
sticky date loaf 62
coconut water 7, 10: coconut &
almond green smoothie 14–15
  orange, ginger & coconut juice 14–15
cod, grilled with radish & herbs 78–9
cookies: oat & sour cherry 86–7
  peanut butter 66–7
corn fritters, smoky 64–5
cranberry & pumpkin seed bars 88–9
cucumber salad & spicy chicken 80–1

D

dahl, creamy red lentil & coconut 26–7
date: date & cocoa energy balls 84–5
  sticky date loaf 62
dried coconut 10

E

egg(s), baked Mexican 40–1
eggplant & salmon miso
skewers 50–1
energy balls, date & cocoa 84–5

F

fennel slaw & turkey burgers 56–7
feta, spinach & mushroom
omelet 42–3

flatbreads, herbed 63
fresh coconut 9
fritters, smoky corn 64–5

G

garlic & cauliflower soup 28
ginger, coconut & orange juice 14–15
granola, nutty 36
guacamole, salsa with vegan chili 52–4

H

harissa & sweet potato soup 38–9
herbed flatbreads 63
horseradish & beet soup 37

I

ice cream, instant 32–3

L

lentil(s), creamy red lentil & coconut
    dahl 26–7
lime: chicken soup with baby
    corn & lime 24–5
    paprika sweet potato wedges 44
    shrimp tacos with coconut 22–3

M

maple bananas with pancakes 68–9
miso salmon & eggplant skewers 50–1
muesli, coconut & berry Bircher 18–19
muffins, orange & poppy seed 70–1
mushroom, feta & spinach
    omelet 42–3

N

noodles (brown rice) & Asian greens
    with shrimp 48–9
nutty granola 36

O

oat & sour cherry cookies 86–7
omelet, mushroom, feta &
    spinach 42–3
opening coconuts 9
orange: orange, ginger & coconut
    juice 14–15
    orange & poppy seed muffins 70–1
organic produce 10

P

pancakes with maple bananas 68–9
paprika & lime sweet potato
    wedges 44
peach & vanilla chia puddings 29
peanut butter cookies 66–7
pearl barley & squash risotto 46–7
pineapple: coconut creams & grilled
    pineapple 30–1
    pineapple carpaccio 82–3
pistachio brownies 72–3
poppy seed & orange muffins 70–1
pumpkin seed & cranberry bars 88–9

R

radish & herbs with grilled cod 78–9
ripeness 9
risotto, pearl barley & squash 46–7

S

salads: grilled chicken with
    Greek salad 55
    spicy chicken & cucumber salad 80–1
salmon, miso salmon & eggplant
    skewers 50–1
salsa 64–5: vegan chili with
    guacamole & salsa 52–4
sauce, slow-roasted tomato 45
seeded coconut flour bread 60–1

shredded coconut 10, 80, 85, 87–8, 91
shrimp: shrimp tacos with coconut &
    lime 22–3
    shrimp with Asian greens &
        noodles 48–9
skewers, miso salmon &
    eggplant 50–1
slaw, fennel 56–7
smoothies: banana & berry smoothie
    bowl 20–1
    coconut & almond green
        smoothie 14–15
soup: beet & horseradish soup 37
    cauliflower & garlic soup 28
    chicken soup with baby corn &
        lime 24–5
    sweet potato & harissa soup 38–9
spinach, mushroom & feta
    omelet 42–3
sticky date loaf 62
storing coconut 9
superfoods 7
sweet potato: paprika & lime sweet
    potato wedges 44
    sweet potato & harissa soup 38–9

T

tacos, shrimp tacos with coconut &
    lime 22–3
tomato, slow-roasted tomato sauce 45
turkey burgers & fennel slaw 56–7

V

vanilla & peach chia puddings 29
vegan chili with guacamole &
    salsa 52–4
vegetables, grilled 76–7

ACKNOWLEDGMENTS

A big thank you to all at Kyle Books, and in particular Claire Rogers. Thank you to Clare Winfield for the beautiful photography, and to Wei for the gorgeous props; it was great to work with such a fun and creative team!

Massive thanks to Nicola for being my right-hand woman and source of much laughter on the shoot days. Also to Poppy for your thorough recipe testing and all-round brilliance.

Enormous thanks to my hugely talented friend, Jenni Desmond, for the wonderful illustrations. I'm so happy that we finally got to work together!

Thank you to my lovely friends, family, and Bob for your support, and for being such eager recipe tasters.

For my mum and Jess

Published in 2016 by Kyle Books
www.kylebooks.com

Distributed by National Book Network
4501 Forbes Blvd, Suite 200,
Lanham, MD 20706
Phone: (800) 462-6420
Fax: (800) 338-4550
customercare@nbnbooks.com

10 9 8 7 6 5 4 3 2

ISBN 978-1-909487-57-4

Project Editor: Claire Rogers
Copy Editor: Eve Pertile
Designer: Helen Bratby
Photographer: Clare Winfield
Illustrator: Jenni Desmond
Food Stylist: Emily Jonzen
Prop Stylist: Wei Tang
Production: Nic Jones and Gemma John

Library of Congress Control Number:
2016939829

Color reproduction by ALTA London
Printed and bound in China by C&C Offset
Printing Co., Ltd.

* Note: all eggs are free-range

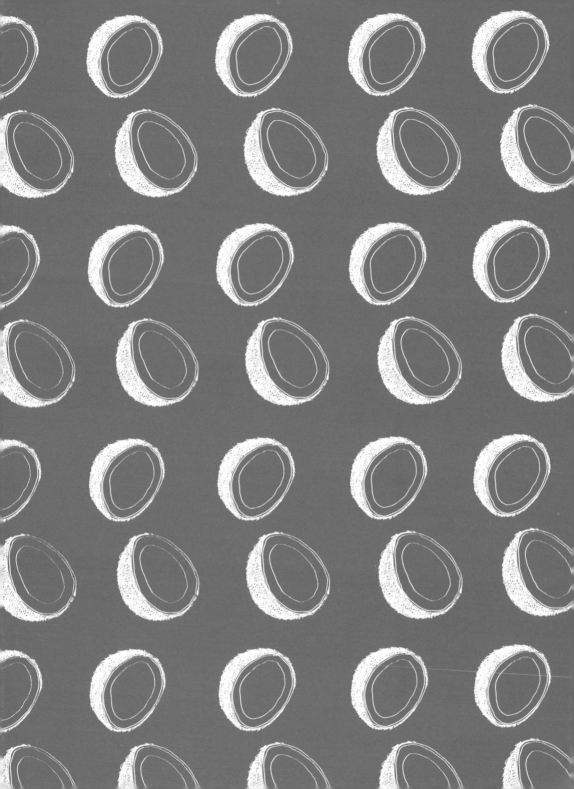